beneath
BLUE WATERS
Meetings with Remarkable
Deep-Sea Creatures

Deborah Kovacs and Kate Madin

Principal Photography by Larry Madin

VIKING

VIKING
Published by the Penguin Group
Penguin Books USA Inc., 375 Hudson Street, New York, New York 10014, U.S.A.
Penguin Books Ltd, 27 Wrights Lane, London W8 5TZ, England
Penguin Books Australia Ltd, Ringwood, Victoria, Australia
Penguin Books Canada Ltd, 10 Alcorn Avenue, Toronto, Ontario, Canada M4V 3B2
Penguin Books (N.Z.) Ltd, 182-190 Wairau Road, Auckland 10, New Zealand

Penguin Books Ltd, Registered Offices: Harmondsworth, Middlesex, England

First published in 1996 by Viking, a division of Penguin Books USA Inc.

1 3 5 7 9 10 8 6 4 2

The authors gratefully acknowledge the help of Shelley Lauzon, Senior News Officer at Woods Hole Oceanographic
Institution, who verified the information on *Alvin*; and Larry Madin, who provided eyewitness accounts of deep-sea dives.

The authors also gratefully acknowledge the photographers who granted permission to
reproduce the copyrighted photographs which appear on the following pages:
Page 20: copyright © H. W. Pratt
Page 29: copyright © Richard E. Young
Page 31: courtesy of Harbor Branch Oceanographic Institution, Inc., Fort Pierce, Florida
Pages 48 and 49: courtesy of Woods Hole Oceanographic Institution,
Woods Hole, Massachusetts. Photographs by Rod Catanach
Page 51: (left) copyright © Bruce Robison; (right) copyright © Bruce Robison, 1988
Page 56 (all): courtesy of Woods Hole Oceanographic Institution, Woods Hole, Massachusetts. Photographs by Larry Madin
Page 57 (both): copyright © Kim Reisenbichler
All other photographs are copyright © Larry Madin.

LIBRARY OF CONGRESS CATALOGING-IN-PUBLICATION DATA
Kovacs, Deborah.
Beneath blue waters : meetings with remarkable creatures / by Deborah Kovacs and Kate Madin   p.   cm.
Summary : Hypothetical scientific tours of the ocean depths present information on the mysterious creatures that live there.
ISBN 0-670-85653-3
1. Marine animals—Juvenile literature. 2. Marine ecology—Juvenile literature.
[1. Marine animals. 2. Marine ecology. 3. Ecology.] I. Madin, Kate. II. Title.
QL122.2.K68 1996   591.92—dc20   96-17755 CIP AC

Manufactured in China
Set in Minister

contents

To the blue-water explorers
— D. K. and K. M.

"Curiouser and curiouser!"

—Lewis Carroll, *Alice's Adventures in Wonderland*

introduction

There is a vast, hidden world on our planet. It lies within the oceans that cover three-quarters of our earth. Its landscape is as strange as that of an alien world. There are pitch-black trenches deeper than the highest mountain peak; towering chimneys that spew clouds of toxic chemicals heated by the earth's core; a 30,000-mile-long mountain range that snakes around the globe like the stitching on a baseball. And less than one percent of this amazing world has ever been seen by man.

The creatures that live here seem as alien as the landscape. Here, in a world too deep for the sun's rays to reach, there live animals whose pulsing bodies send out ribbons of light. A glow-in-the-dark fish lures its prey with a fishing rod attached to its head. An animal that looks like a drifting trash bag waits for a meal to swim inside the bag, then cinches itself shut. A creature whose name means "vampire squid from hell" creates its own light with light organs—some lidded like eyes—scattered over its body. For millions of years, these animals and countless others have lived, fed, bred, fought, and fled, their existence unknown to those outside their chilly, watery world.

Meanwhile, seemingly a world away but sharing the same planet, live humans with their built-in curiosity. Among those who are most curious about the living world are scientists called naturalists. One of the greatest naturalists was Charles Darwin, who lived in the first half of the nineteenth century. Darwin's curiosity led him to take a four-year voyage aboard the H.M.S. *Beagle,* observing creatures of land, sea, and air. He spent the rest of his life studying what he had found and wondering how each living thing makes its way in the world.

More than one hundred years after Darwin's death, naturalists are still asking the same questions about the environments they study: What lives here? Has this creature ever been seen by man before? How does it feed and find a mate? Like

Darwin, naturalists today are also taking long voyages of discovery. And today they can visit worlds that Darwin never dreamed of.

To make these journeys, naturalists who study life in the oceans—scientists called biological oceanographers—must prepare to enter an environment very different from our own: airless, cold, virtually gravity free, dark even in the daytime; a place where it's easy to lose even your sense of up and down, and where the pressure can be more crushing than a steamroller. Like astronauts who rely on spaceships and space suits to survive during their explorations, biological oceanographers must rely on ships, scuba gear, and research submersibles to explore the deep.

The ocean environment changes the further one travels down the water column, the term oceanographers use to refer to ocean waters from the surface down to the seafloor. The types of creatures one sees in different levels of the water column also change, for survival becomes much harder the deeper one goes. And survival is harder for the scientists visiting this world, too. As they dive deeper, the water becomes colder, the water pressure greater, and the dark darker, until light is only a memory.

In this book you will travel with biological oceanographers more than a mile down through the water column, from the sunlit surface of the open ocean to the ocean floor itself. In three different dives you will explore life in three of the ocean environments which make up the water column. Scuba divers will guide you through the level closest to the surface, where the waters are still lit by the sun's rays. Then you will join scientists who travel in a submersible to the midwaters, a place high above the seafloor but completely dark to human eyes. Finally you'll dive with the crew of a special deep-sea submersible to the area above the seafloor. Since in real life no single dive would show you all the wonders of each level, we have combined events from several dives. Though these are hypothetical dives, everything that happens in each dive once happened to a working oceanographer.

It is only in the last twenty-five years that biologists have been able to study the creatures of the deep, open ocean in their natural habitat. Only about fifty scientists worldwide have been involved in this type of research. Such a small part of this world has been explored that every research trip holds out the chance to discover a new creature. Sometimes scientists try to capture one of these new and amazing creatures and bring it back to the surface. But once it is brought into our world, the animal loses the luster and mystery it had in its own world. Our home is just as alien to it as its environment is to us.

This undersea world is the last frontier on earth, the one place not explored and settled by humans. When you journey with the scientific adventurers into the deep, remember the natural curiosity that drives them. As Charles Darwin wrote at the end of his life: "It has been said that the love of the chase is an inherent delight in man—a relic of an ancient passion. . . . If, as the ancients supposed, the flat earth was surrounded by an impassable breadth of water, or by deserts heated to intolerable excess, who would not look at these last boundaries to man's knowledge with deep but ill-defined sensations?"

Cross-Section of the Water Column

To help study and discuss the ocean, the largest part of the earth, oceanographers have divided it into different areas called "zones," based on depth, light, and what lives there. Our three dives take place in the "open sea" or "blue water," the part of the ocean that is away from the shore and has an average depth of 3 1/2 miles (5 3/4 kilometers). Seawater near the shore contains sediments churned up from the bottom and silt and chemicals that flow in from rivers. Water in the open ocean doesn't have all these particles in it and is much clearer, which means the rays of the sun can travel deeper. To living organisms, this light is perhaps the most important thing that changes with depth.

It is important to remember that these zones are not hard and fast divisions. There are no walls in the sea. Water and the creatures that live in it are constantly moving. An animal that spends most of its time in the mesopelagic zone may also feed in the epipelagic zone, and vice versa. Classifying the animals by zones simply gives researchers a better idea of where each creature is most likely to be found.

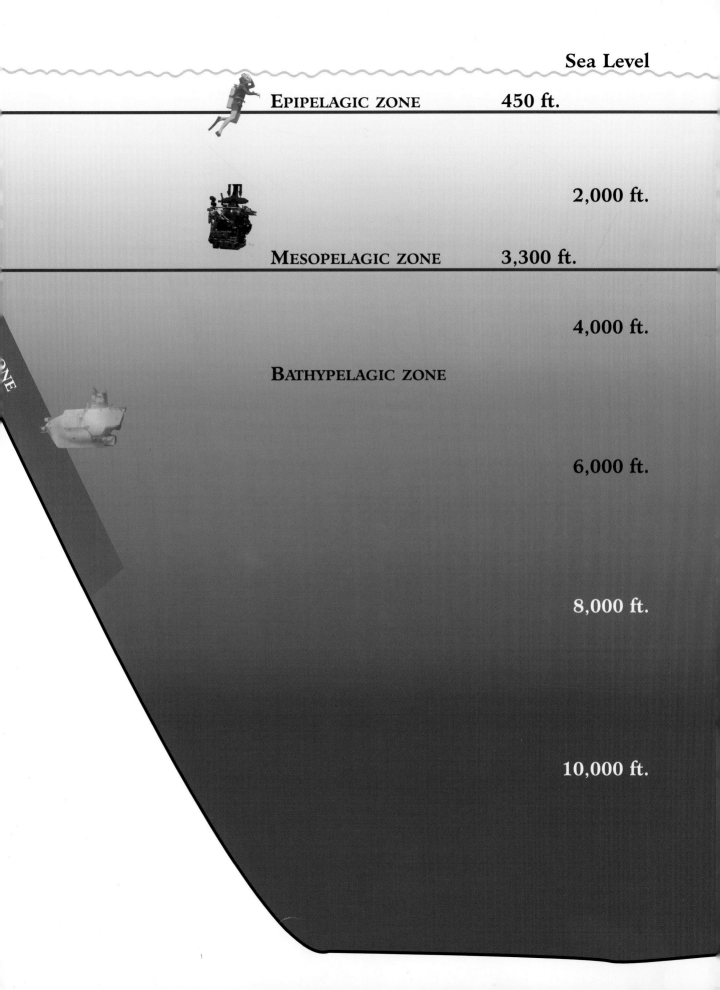

Sea Level

EPIPELAGIC ZONE   450 ft.

2,000 ft.

MESOPELAGIC ZONE   3,300 ft.

4,000 ft.

BATHYPELAGIC ZONE

6,000 ft.

8,000 ft.

10,000 ft.

The Epipelagic Zone

Two things define the uppermost part of the water column, the epipelagic zone: the light of the sun and the crystal blueness of the water. The epipelagic zone reaches from the surface to about 450 feet (150 meters) below the surface. This is the part of the water column where plants and the creatures that feed on them live. In the open ocean, the sun's rays penetrate to about this depth, allowing a wide variety of plants to grow, including the smallest but most important plants in the sea, the microscopic phytoplankton. Small animals called zooplankton eat the phytoplankton, and they in turn are eaten by larger animals. All these creatures are linked together in a food chain, from the tiniest plant to the largest fish and marine mammals. With food for many types of creatures in such abundance, the epipelagic zone is the richest, most life-filled part of the ocean.

In the open ocean, which is about 100 miles (160 kilometers) from shore and 2 miles (3 kilometers) above the chill, rocky ocean floor, the waters of the epipelagic zone are an incredible, beautiful, pure blue. These waters are too far away from the shore to be affected by such things as eroding soil or river runoff. As opposed to shallower coastal water, this "blue water" does not have many particles floating in it. Even on a cloudy day, the water is the most crystalline blue imaginable. This zone has much to study, and perfect conditions to study in.

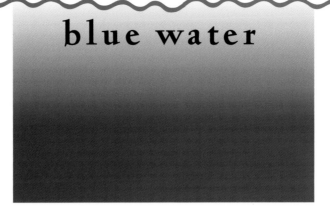

blue water

Sixty feet beneath the surface, in the vast blueness of the open ocean, two creatures swim in an utterly blue world. Nothing but water is visible in all directions. Beams of sunlight slant through the light blue water at the surface, shimmer down onto the creatures' backs, and shoot past them into the dark blue below, to finally disappear into the pitch black that hides the lower depths. One of these creatures, a dolphinfish, is at home in this world. The other is alien to the ocean, and must mimic the dolphinfish so that he can visit.

The dolphinfish is perfectly adapted to its world. Very sensitive vibration detectors along its sides allow it to "feel" what is happening in the water around it. Its eyes are adapted to the way light travels through water, rather than through air like land creatures' eyes. It is cold-blooded, as are most of the creatures of the sea. There is no air to breathe here, but the dolphinfish does not need air. It takes in the oxygen it needs directly from the water passing across its gills.

The other creature is a human diver. He is not built to live in this world, so his survival depends on many pieces of special equipment. In order to breathe, he wears an

A dolphinfish, Coryphaena (kor-ee-FEE-na), about 5 feet (1 1/2 meters) long. Most large fish avoid divers, but this one was unusually curious.

air tank strapped to his back and connected to a rubber hose and a regulator that he holds in his mouth. He wears a face mask to protect his eyes from the salty water and let him focus clearly on what's around him. Humans are warm-blooded, so the diver wears a rubber wet suit to keep him from losing body heat to the water.

The diver needs equipment not just for survival but in order to move around in the water. For efficient swimming, feet that normally walk on land must be covered with flippers, which are used in the same way a fish uses its fins. To help him sink below the surface to observe creatures like the dolphinfish, he wears a weight belt. When he wants to rise, he inflates a vest called a buoyancy compensator (B-C for short). By adding air to the vest, he can drift upward. The dolphinfish has a similar ability. Its body contains a gas bladder in which it can increase or lessen the amount of gas, in order to help itself rise and sink in the water.

A blue-water diver, a sample jar held at the ready, hangs on to his tether.

The dolphinfish can travel through miles of open ocean on its own. But the diver would not be safe by himself. In the open ocean it's easy for a diver to lose track of where he is, and even of which way is up. The sea surface is far overhead and the seafloor is too far down to see. There are no landmarks to guide him, just unending blue. If the diver becomes disoriented, he may not be able to reach the surface when the air in his tank runs low. So this diver is tethered to four others, one of whom is a safety diver. The safety diver sits in the middle of the tethers, like a spider in its web, watching all the other divers and making sure that the waters around them are safe. Connected together, the safety diver and the four research divers who make up the team have come to the dolphinfish's world to learn more about the creatures that live here.

Research divers don't swim together just for safety. The sea is so vast, and the creatures in it often so far apart, that the more researchers who are out looking, the more likely they are to find something. This team is particularly interested in invertebrates: animals without backbones.

The safety diver holds the down line, which is attached to a buoy at the surface. Divers can check the down line to see how deep they have swum, while the crew on the surface watches the buoy to keep track of the divers' position.

Most of the animals that live in the oceans are invertebrates; the great variety of their form and function and their perfect adaptation to this watery world make them fascinating. Though the animals with backbones, such as the dolphinfish, are not what the team has come to study, the diver knows how unusual it is to be face to face with this fast-moving fish in its own element. Moving slowly so as not to stir up the water—for all creatures of the sea are sensitive to the slightest shifts in the water's flow—the diver unclips the camera from his dive vest and takes a picture of this special sighting.

Meanwhile, another member of the team is searching for salps, transparent jelly-like animals with tube-shaped bodies. The salp swims by constricting and relaxing bands of muscles which pump water through its body, drawing the water in at the front and expelling it at the rear. With each expulsion of water, the salp shoots forward. As it swims, it eats, using a sticky mucus net suspended within its body to filter out small particles of food from the water that passes through.

The diver hopes to collect some of these creatures so she can study how the salp's body works: what it eats, how long it takes to digest its food, and how much oxygen it uses. She opens her eyes wide and stares into the blue water. An individual salp is about the size of a nickle and almost completely transparent—not an easy thing to see. But the diver is patient and knows that salps are rarely found alone; either they live as individual members of a large group or they are born connected in a chain. Because a salp's body is asymmetrical (not the same on both sides), it tends to travel in a circle when it makes its pulsing swimming motion. Salps swim more efficiently and thus find more food when connected to other salps.

Soon the diver spots something that looks like a very long, pulsing, dotted line looming up out of the endless blue. It is about as long as a school bus and contains hundreds of individual salps, each one virtually transparent except for its opaque stomach. A fellow diver also notices the salp chain and signals that she's going to help with the collecting. Both divers tug sharply at their lines to let the safety diver know that they'll be swimming to the end of their tethers. She gives the OK sign and watches them carefully, making sure their lines stay free of tangles.

The divers swiftly flipper over to the salp chain, grabbing collecting jars from the pouches around their waists and opening them as they swim. Before the dive, these jars were filled completely with sea water, and their tops securely tightened. When their tops are removed, the filled jars won't give off air bubbles, which could easily scare an animal off. Also, a jar filled with air would be impossible to open, because, even a few feet deep, the water pressure on the outside of the jar, that is, the weight of the water pressing down on the jar, would force the lid down too tightly.

The divers reach for the chain with their open collecting jars. Each manages to detach and collect a few of the creatures. After giving each other the OK sign, the divers return their collecting jars to their pouches and swim back closer to the safety diver.

On the other side of the safety diver, the diver who photographed the dolphinfish is now carefully observing *Geryonia proboscidalis,* a jellyfish whose name literally means "big-nosed three-bellied monster." *Geryonia* does not actually have a nose—or a face, or even a head. The "nose" part of its name refers to the elephant's-

A 20-foot- (7-meter-) long Salpa maxima (SALP-ah MAX-ih-muh) chain, above.
Below is a close-up of two individual salps in the chain. Salps are part of a
group of animals called tunicates, which are closely related to vertebrates.

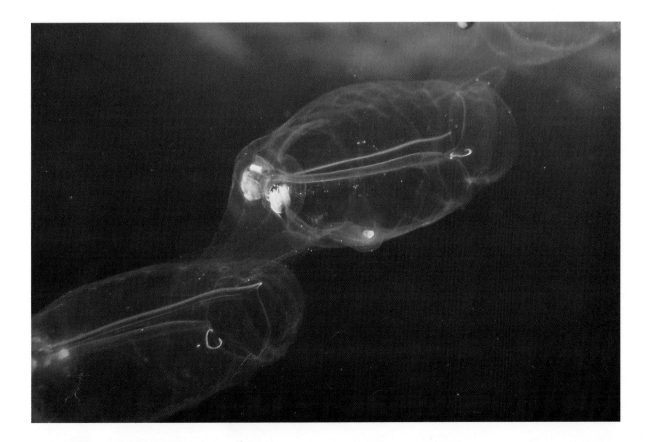

What's in a Name?
A Note on Scientific Naming

With so much of the sea being charted for the first time, new organisms are frequently being discovered. How do the people who discover these creatures name them so that scientists all over the world will be able to recognize them—even though they've never actually seen them?

This question was first addressed in the 1600s when the microscope came into use. As scientists began to explore the vast miniature world of plants and animals, they realized that there were many more types of living things on earth than they had ever suspected. There was such a rush of new discoveries that the information seemed overwhelming. People in different countries had different names for the same animals and plants, adding to the confusion.

Everything got sorted out when Swedish naturalist Carolus Linnaeus published his book Systema naturae in 1758. He created a system for naming living things in Greek and Latin, the standard languages of scholarly works in the eighteenth and nineteenth centuries, and scientists today continue to follow this tradition. The system outlines how to classify every known plant and animal by sorting them into categories that show how each creature is related to others, what it has in common with them, and what makes it unique. He started by dividing all creatures into very general categories, then broke these into increasingly more specific categories, until finally only two words were needed to identify any living plant or animal. His system is called binomial nomenclature, which means two-part naming.

The first part of the name is the genus, which describes a large category of creatures. For example, all bears are in the genus Ursus, Latin for "bear." The second part of the name, the species, identifies what specific type of bear it is. The black bear of America is thus named Ursus americanus, which distinguishes it from other bears such as Ursus arctos (the brown bear of Europe).

A black bear is a fairly well-known animal, so its common name is the one most often used to describe it. Some of the creatures in this book have a common name, too. When they do, we'll provide it, along with the scientific name. But many of these animals are so rarely seen that they have no common names. Some are so newly discovered that they don't even have a species name yet.

16

When a new animal is found and classified, how is a name chosen? Most researchers name the creature after some obvious physical feature or behavior. For example, a shrimp which made loud snapping noises by frequently clicking its claws was named Oratosquilla, which means "speaking shrimp." Sometimes an animal is named after a respected person, a friend, or the person who discovered the creature. Though naming a new animal is a serious business, sometimes scientists have some fun with the process. One scientist may have named a new jellyfish Lizzia blondina for a blond-haired woman named Elizabeth!

Oratosquilla investigatoris (oh-ra-toe-SQUILL-ah in-ves-ti-ga-TOR-iss).

Geryonia proboscidalis *(jair-ee-OH-nee-uh prob-ah-sid-AL-is), on the prowl for a meal.*

trunklike tube that is the creature's mouth and stomach. The diver hopes to see the jellyfish use this appendage to capture a meal. He knows that all jellyfish are constantly on the prowl for food, so he simply holds his camera steady and waits for the jellyfish to find its prey.

Floating still in the water, the jellyfish seems to be just a pale blob—it has no central nervous system and few distinguishing features. But when it moves, it becomes pure grace. Muscle fibers around the rim of the jelly-filled part of its body (which is called the "bell" or "umbrella") contract, causing the jellyfish's body cavity to eject water. This sends the animal shooting forward at a remarkable speed, much the way the salp moves. When the jellyfish is at rest, the umbrella relaxes and the creature looks like a gently descending parachute.

The diver now watches as the *Geryonia* comes upon its prey, a crustacean related to the lobster that is no bigger than a finger-nail. It brushes the crustacean with the long tentacles that descend from its bell. These tentacles are lined with stinging cells called nematocysts, which instantly stun the jellyfish's prey, making it unable to move. Now the *Geryonia* draws the creature up through its dangling mouth and into its simple stomach. Mission accomplished—for the hungry jellyfish and for the diver who captures the hunt on film.

Another diver is watching the glittering approach of *Eurham-phaea vexilligera*, a creature that looks like a neon jellyfish. But it

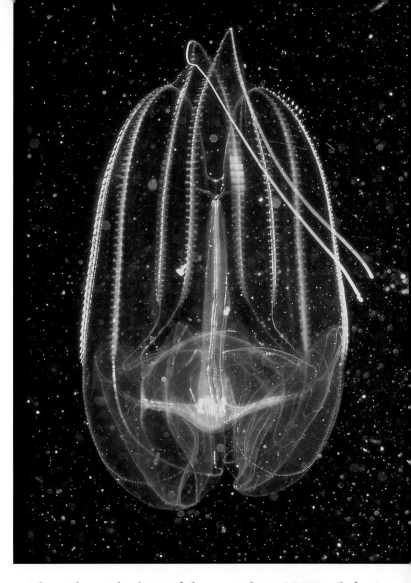

The eight comb-plates of the ctenophore (TEEN-oh-fore) Eurhamphaea vexilligera (your-ham-FEE-uh vex-ill-IH-jer-uh) can clearly be seen as they refract the light from the surface. The animal's bottom lobes are covered with sticky tentacles, which help move food into its stomach.

isn't a true jellyfish—and even the light it flashes is an illusion. The *Eurhamphaea* is a comb jelly or ctenophore, named for the eight rows of comb-plates that sur-round the creature. A comb-plate is just what it sounds like: a set of filaments, called cilia, which are arranged side by side like the teeth on a comb; these combs are arranged in long rows along the animal's body. When the combs are moved in wave patterns, they act like tiny oars, rowing the ctenophore through the water so it can search for food. Ctenophores are the only animals in the world that move this way or that put on

19

Now You See Me, Now You Don't:
Camouflage, Protective Coloration, and Transparency

Many ocean predators use their eyes to search for prey. Their prey is dependent for its survival on being able to either outswim the predator or hide from it. In shallow waters, near the shore, or over reefs, there are many hiding places and many ways to hide. Many sea animals here have colors and patterns on their bodies that mimic or blend in with their environment, giving them a natural camouflage. Animals that live above the seafloor in shallow waters may have mottled coloration that looks like light and shadow playing on the sand; others may have brown or gray coloring to blend in with rock formations; others, such as the common octopus, can change their skin color or pattern to match the environment around them.

In the open ocean, however, there is nowhere to hide—no reef, rock, or sandy seafloor. Animals living here have evolved different kinds of protective coloration. Most epipelagic vertebrates—fish, sharks, dolphins, and whales—have a coloring called countershading. Their stomachs are white, which makes it hard for a predator looking up into the light to see them. Their backs are dark blue, gray, or black, so that seen from

A tuna with countershading.

above, they blend into the deep blue-to-black of the ocean depths below. Some have silvery sides, reflecting the water and light level around them. If a countershaded fish sees a predator approaching, it has at least a chance of swimming away before it's spotted.

Many zooplankton can't swim fast enough either to escape a predator or to overtake prey. Some can't swim at all, but drift wherever the currents carry them. The survival of these zooplankton—both predator and prey—depends on their invisibility. They are aided in this by having very simple structures: their bodies are thin and made up mostly of water; they have no complex internal organs, and no bones. Some zooplankton are transparent, as difficult to see as a sliver of clear glass in the water. Others are a transparent blue, like pieces of solidified water. Others are clear except for one or two plainly visible body parts, which can be opaque, dark, or brightly colored. Scientists aren't sure why some animals have evolved this type of coloration. The opaque parts could be an attempt to mimic another animal and confuse predators. They could be a kind of decoy, since predators have been observed to attack and eat only the visible part of prey. Of course, some parts of an animal, such as the stomach, are almost impossible to make transparent. A transparent body can't hide an animal's nontransparent lunch!

The immature forms of some crustaceans, such as crabs and lobsters, are completely transparent. This gives them a much better chance of survival during their trip down through the ocean, before they settle on the bottom and develop the hard outer armor of adulthood. This transformation can happen amazingly quickly, with some larval crustaceans getting their hard shells in just a few hours.

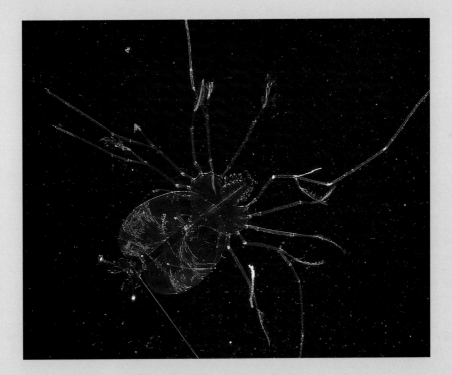

The larval form of the spiny lobster, *Phyllosoma*
(fill-oh-SOHM-ah), has a body as flat as a leaf,
hence its name, which means "leaf-body."

The oceanic white-tipped shark, Carcharinus longimanus
(car-car-EYE-nus lon-gee-MAN-us), is one of the most ancient types of fish.

such a show as they move, throwing off bright fragments of prismatic color from their rippling combs. The effect is startling, but the sparkles are simply surface light refracted through the evenly spaced teeth of the combs, in the same way that when you squint at a bright light, your own evenly spaced eyelashes sometimes break the light into the colors of a rainbow.

Comb jellies are also different from jellyfish because they do not have sting-ing cells on their tentacles. Instead they have sticky "glue cells" with which they catch their prey.

Eurhamphaea is special among comb jellies because it secretes a luminescent ink-like liquid when it is disturbed. The diver wants to capture this creature so she can study the "ink" it produces. Moving slowly and carefully, she opens a collect-ing jar, and heads toward the animal—when she feels a strong pull on her safety tether.

She looks up to see the safety diver pointing down. Something is moving up toward the divers . . . something very large and possibly very hungry: an oceanic white-tipped shark.

The shark, an aggressive meat-eater, begins to circle the divers. But the divers are experienced and do not panic. Instead, they draw closely together. The divers know that, though sharks are capable of seriously injuring, even killing a swimmer, they rarely do. And they are far more likely to attack a lone, distressed diver than people diving close together. This shark will probably keep circling the group warily for some time, then move on. While she waits with the others, the diver who lost the *Eurhamphaea* uses the time to take a picture of the shark, something few people have the opportunity—or the desire—to do.

When the shark swims away, the divers quickly spread out again. Their time underwater is limited by the air in their tanks to forty-five minutes, only fifteen minutes of which are left.

Another diver interested in ctenophores spots a *Beroe cucumis* drifting by. It looks like its namesake, a cucumber, but acts like a swimming stomach. The *Beroe* hunts other comb jellies by swimming until it literally bumps into its meal. It then opens its huge mouth and swallows the un-lucky find whole. It can hold the prey in its mouth—or even bite off a piece—with the fused cilia on its lips, groups of hairlike projections which act as teeth.

The diver knows the *Beroe* eats other comb jellies but wants to find out which ones. He tries to collect it in a jar, but the jar slips from his grasp and slowly falls and falls. When something falls, it's natural to

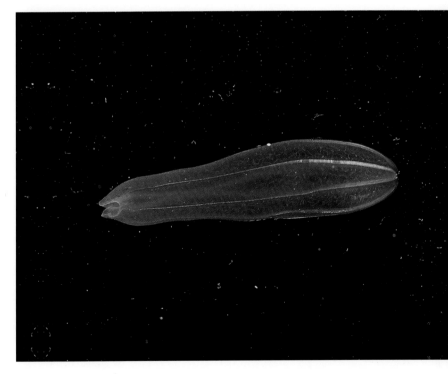

Beroe cucumis *(bear-ROW-ee kyu-KYU-miss).*

Carinaria richardi (ker-ih-NAR-ee-uh rich-ARH-dye) is related to the garden snail. But unlike snails that live on land or by the shore, Carinaria *has only a tiny shell.*

try to catch it, and the diver instinctively reaches out for the jar while swimming down to the end of his tether. But it could be fatal to go after the jar. He can only watch it fall to the bottom of the down line, and keep falling down and down. He watches until the jar disappears into the inky blackness of the deep sea.

His disappointment is short-lived, for soon he spots an interesting snail called *Carinaria richardi,* an animal with a big appetite, a long snout, and beady eyes that remind the diver of a dog. He knows that another diver on the team is studying this little creature, so he collects it, this time keeping a firm grip on the jar.

Closer to the surface, the safety diver is so busy watching the others that she doesn't notice the swift approach of a *Diphyes dispar.* Though it looks and acts like a single jellyfish, it is actually a siphonophore, a composite of jellyfishlike creatures that function together like a single creature. As it passes, its trailing tentacles brush her hand, which she suddenly draws back in pain, for it feels like she's just been stung by a bee. She quickly turns and recognizes the *Diphyes,* and realizes that she's been stung by its nematocysts. Though her hand is quickly turning numb, she knows that she will soon regain feeling in it and that the sting is not serious, which is not always the case with jellyfish and siphonophore stings. The sting of the cubomedusa or "sea wasp," for example, can cause severe pain and, in some cases, death.

Flexing her numb fingers, the safety diver checks her watch. Dive time is over. She tugs the tethers to get the other divers' attention, then points to her watch, then up to the surface. They understand that the day's expedition is over. As the divers swim toward the safety diver, they pass two more jellyfish. They don't have time to stop for samples and observations, but they can admire the beautiful creatures.

The yellow insides of *Leuckartiara octona* make the creature stand out in this blue realm. *Porpita porpita* is fairly common in tropical surface waters. Though this is not a rare sighting, it is a beautiful one; this relative of the jellyfish shimmers like

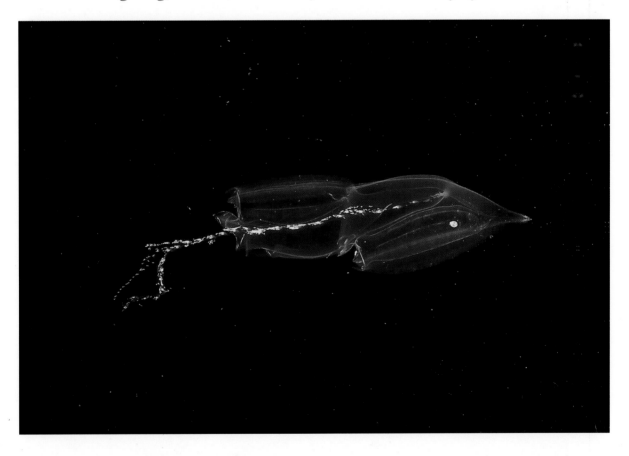

A Diphyes dispar (dye-FYE-eez DISS-par) swims by with its tentacles contracted. The shiny silver dots on its tentacles are clusters of nematocysts, groups of stinging cells that shoot out tiny barbs at anything that comes in contact with them.

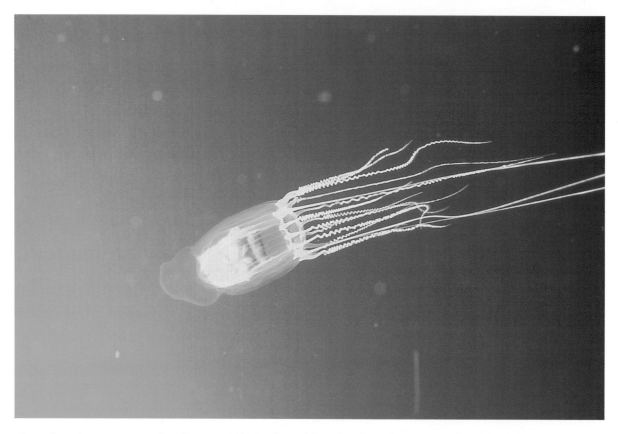

Leuckartiara octona *(loo-kar-tee-EHR-uh oc-TONE-uh), above, stands out in the ocean with its bright yellow stomach. The blue color of* Porpita porpita *(por-PEET-uh por-PEET-uh), below, helps this 3 inch (5 centimeter) creature blend in with its surroundings.*

a chrysanthemum-shaped firework that will never go out.

When all the divers have reached the safety diver, they unclip their tethers and swim slowly upward. But just as they poke their heads above water, the boatman calls to them to look out for Portuguese men-o'-war. Several of the stinging siphonophores are very close to the port side of the boat! The divers quickly swim around to the other side of the little inflatable boat and clamber rapidly and clumsily over the side.

Safely in the boat, one scientist leans over and slowly and carefully scoops up one of the men-o'-war into a wide-mouthed container, to take it back to the shipboard aquariums. No one is stung—this time. Now they can look forward to a hot shower and a hot meal aboard ship, after their preliminary examination, listing, and logging in of the animals they have collected.

The glassy, transparent float of the siphonophore Physalia physalis *(fi-SAY-lee-uh fi-SAY-lis), or Portuguese man-o'-war, is easy to spot on the surface of the water. Its blue tentacles, which can be as long as 30 feet (10 meters), are far harder for a swimmer or a fish to see. The man-o'-war's sting paralyzes its prey and can cause humans severe pain lasting for days, as well as red welts and swelling.*

The Mesopelagic Zone

Below the sunlit regions of the epipelagic zone is the mesopelagic zone. This is the "twilight zone" of the ocean, a region of deepening darkness, ending around 3,300 feet (1,000 meters) below the surface, a depth where it is always night. Below 2,300 feet (700 meters), human eyes see only darkness, though many of the animals that live in this zone have eyes strong enough that they can still tell night from day.

Only one hundred or so people have ever visited this region, for only twenty research submersibles are capable of carrying people to these depths, where the water pressure is fifteen to one hundred times greater than the air pressure on land.

Life is hard for the creatures that live here. They have to spend most of their time drifting or cruising in search of food, using up as little energy as possible in the process. There are no plants here, because not enough light filters through to support photosynthesis, the process by which plants use the sun's energy to produce food. With no plant life, almost every creature is both predator and prey. The creatures that live here are scarce and hard to see—which makes the scientists' discoveries all the more rewarding.

down into
darkness

In the dark entrance to the mesopelagic zone, at a depth of about 450 feet (150 meters), a squid called *Heteroteuthis* zips through the water. Contracting its muscles sharply to expel water from its body, it thrusts itself forward and soon overtakes the shrimp that is its prey. Though now the hunter, *Heteroteuthis* may at any time become the hunted. If attacked or alarmed, this squid, like most other squid, will discharge ink to confuse the predator. However, this little squid, no longer than a pinky finger, has a special ability. To work effectively in a world without light, its ink glows.

Heteroteuthis *(het-er-o-TOOTH-iss).*

Johnson Sea-Link *being lowered by crane from the deck of its mother ship.*

Squid like *Heteroteuthis* have lived in midocean waters for tens of millions of years. Though they are sometimes caught in fishing nets and hauled to the surface, people rarely get to observe them in their own world. But the squid that just caught lunch is being closely watched, photographed, and videotaped by a scientist and a pilot who peer out through the Plexiglas bubble of the research submersible *Johnson Sea-Link,* called *Sea-Link* for short. They, along with a technician who serves as a backup for the pilot and another scientist who crouches by a porthole in the rear of the sub, have just five hours to spend exploring these deep, dark waters. The purpose of the mission is as simple to state as it is difficult to achieve: to spot, observe the behavior of, and in some cases collect as many midwater invertebrates as they can.

The scientists have many questions they want answered: What kinds of creatures live at this depth? How do they find their prey? How do they avoid predators? What do they do during the day? During the night? Can they even distinguish day from night at this depth? But the scientists have far, far more questions than any one dive—or even several—can answer. Far fewer creatures live in this midwater than in the zone above, and finding anything is simply a matter of luck. It's impossible to predict what they will see, or if they will see anything at all.

DOWN INTO DARKNESS

The sub has descended quickly from the surface to the midocean world and will now travel slowly down through the mesopelagic zone, from its current depth of about 450 feet (150 meters) to about 3,000 feet (900 meters), pausing to observe any animal the scientists come upon.

The two scientists have different views of the mesopelagic world. The scientist sitting up front is in charge of the dive and gets the best seat. Through the Plexiglas sphere that is the main part of the sub he can scan a wide area and study the larger animals. He has such a clear, open view here that he sometimes feels as though he's actually drifting in the water outside *Sea-Link* along with the animals he has come to observe. The scientist in the rear of the sub has a small porthole just beside her seat, through which she can get a close-up view of small animals. The two compartments are separate, and the crew members keep in touch by intercom. The pilot can maneuver the craft to bring a creature into either viewing point.

No matter how deep the submersible dives, the pressure inside the passenger compartments remains the same as that found at sea level. At a depth of half a mile, the pressure of the water pushing against the sub is more than eighty times greater than the pressure inside the sub—over 1,200 pounds per square inch!

Though the conditions may sound lethal, and the very idea of being crouched inside a little bit of metal and plastic a half-mile down in the ocean sounds improbable, the divers are at surprisingly little risk. The technology of ocean exploration has advanced a great deal in the last half-century. To many of today's oceanographers, a sub dive seems no more dangerous than driving to the store. Research

Sea-Link *in descent.*

Light from Life:
Bioluminescence in the Mesopelagic Zone

On some summer nights, nature puts on a light show in an open field, as fireflies flicker on and off. The open ocean can also put on a show, as thousands of microscopic creatures at the water's surface emit tiny sparkles of light. Sometimes an animal such as a jellyfish or ctenophore will add its much larger but less frequent flashes.

This light show is called bioluminescence (commonly referred to as phosphorescence). Many forms of life can produce light, from tiny bacteria and fungi to larger plants and animals. Most of the world's bioluminescent animals are found in the ocean, including species of jellyfish, ctenophores, worms, crustaceans, clams, squid, octopus, and fish.

Though bioluminescence is commonly seen at the sea's surface, most luminous animals make their homes in the mesopelagic zone. In these midwaters, light flashes produced by living things can even mask the faint sunlight filtering down from above. Most bacteria glow constantly, while most animals flash, sometimes hiding their glowing ability until attacked. There are fish that have special sacs or pockets filled with glowing bacteria, which are covered by tissue like an eyelid. These "eyelids" can be opened or closed, so the fish can flash light just when it wants to. The tunicate Pyrosoma, whose name means "fire-body," glows only when touched or in

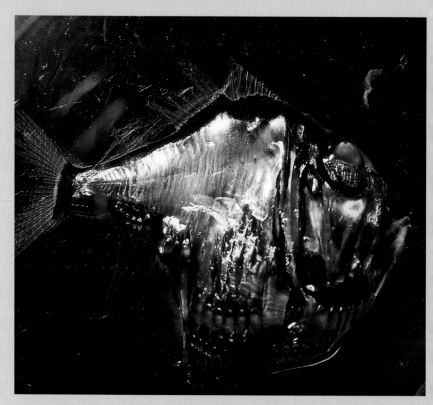

The silver scales of the hatchet fish reflect the light of any bioluminescent creatures near it—or a photographer's flashbulb, seen reflected just below the fish's eye.

response to other light. When a nineteenth-century naturalist caught a colony of these creatures in a net, he discovered that he could actually "see his name in lights" by writing across the colony with his finger.

What is the point of all this light? There are a number of possibilities. Not all creatures use light for the same purpose. Some, like Pyrosoma, only flash when touched or startled by a sudden flash, possibly using their light to startle in turn anything that comes too close. Many bioluminescent animals eject glowing secretions or clouds when attacked. They often dart away at the same moment, using the temporary flash of light to confuse a predator and make good their escape.

Mesopelagic shrimp, squid, and fish often have complex light-producing organs called photophores scattered over their bodies. Photophores may contain lenses to direct the light, reflective screens to enhance it, and even filters to change its color. Males and females of a species may have different patterns of photophores and flash different signals. In the open ocean, where members of one species may be widely scattered, these lights could be used to attract members of the opposite sex—a sort of deep-sea neon advertisement.

As well as luring mates, lights can also lure prey. Like moths to a flame, midwater animals are instinctively drawn toward light. Deep-sea anglerfish have photophores at the end of special spines that extend from their fins. These dangle in front of their mouths, like glowing lures on

The photophore-tipped "fishing rod" of the deep-sea anglerfish, or *Phrynichthys wedli* (frih-NICK-thees WED-lye), floats above its head as it swims. The lines surrounding its head like strings of beads are part of this fish's highly sensitive motion detection system, which helps it find prey in the dark waters.

Dozens of photophores, which look like tiny beads, can be seen scattered over the body of the squid *Histioteuthis* (hiss-tee-oh-TOOTH-iss).

the end of a long fishing pole. Except for this lure, the fish have very dark skin that absorbs light, allowing them to wait unseen behind their "bait."

One of the most amazing uses of bioluminescence in the midwater is as counter-shading, which has been best studied in fish. These fish have large photophores on their stomachs and can control the light they emit to match exactly the intensity and color of the bluish remains of down-welling sunlight around them. At whatever midwater depth the animal finds itself, it can virtually disappear from view to any predator looking up from below.

submersibles are well designed and carefully constructed to ensure their passengers' safety. Before a dive, the crew goes through a series of safety checks to head off any problems. Allowances have been made so that anything that goes wrong in the sub's five-hour dive can be handled by its crew or corrected by the sub's multiple backup systems. Should anything happen to the pilot, for instance, the technician can bring the submersible up from his controls in the back compartment.

Most of the time, the sub travels in darkness, scanning the waters for the flashes and spots of light which are the only indications of life at this depth. The lights come from bioluminescent creatures, animals that produce their own light, and from the mesopelagic fish whose shiny, metallic-looking scales reflect this light.

But *Sea-Link* has lights, too, which the pilot turns on at the scientists' request. The scientist at the rear asks for a light so that she can observe a *Tomopteris* now

Tomopteris *(tom-OP-ter-iss) is a relative of the earthworm,*
and a common sight in the open ocean.

swimming into view. The 4-inch- (10-centimeter-) long worm is using its many paddle-shaped appendages to row through the water. It swims in S-shapes past the porthole, moving like a sidewinder rattlesnake traveling over sand. In the reflection from the sub's light, its transparent body appears to glow.

The scientist up front also requests that his light be turned on, to allow him to watch the constant cascade of white particles drifting down from above. This is marine snow—the litter of the ocean—the tiny remains of plants and animals that once lived in the upper waters. Though they look like snowflakes falling on a moonless, windless night, they are more like tiny packages of sunlight that fall to feed the creatures of the lightless deep waters.

A jellyfish pulses up out of the darkness and into the sub's spotlight, making the scientist in front sit up excitedly. This is *Benthocodon hyalinus*, a creature discov-

Benthocodon hyalinus *(benth-oh-KOH-don high-a-LINE-uss).*

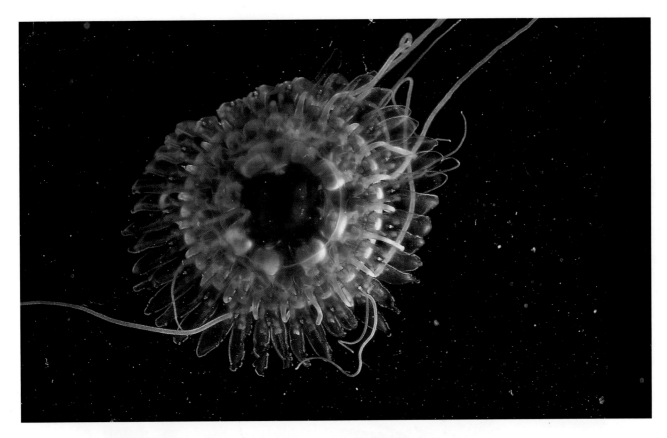

Atolla wyvillei *(uh-TOLE-uh WY-vill-eye)*.

ered in 1987 and seen only about a dozen times since. The jellyfish is transparent except for its stomach, which appears bright red in the light from the sub. Without the sub's light, the stomach would appear black to any of the animals which can see in this dim twilight, for the red part of the spectrum is absorbed by the waters above.

No one knows for sure why this jellyfish and many other mesopelagic animals have red insides. It has been suggested that, since the red color of the stomach sacs absorbs blue light rather than reflecting it, the light from any glow-in-the-dark creature that *Benthocodon* might eat would be concealed in the black-looking sacs. The jellyfish would thus still be invisible, safe from discovery by other predators.

About twenty minutes later, the scientist in the rear sees another exciting jellyfish, *Atolla wyvillei*. It pulses through the water past her small porthole, its tentacles waving and its bright red and orange tissues glowing in the sub's light. Without the artificial lights, it would appear black at this depth, like the *Benthocodon*. But to the sub's crew, it looks brilliantly, shockingly colorful, like a cross between a jellyfish and a sunflower.

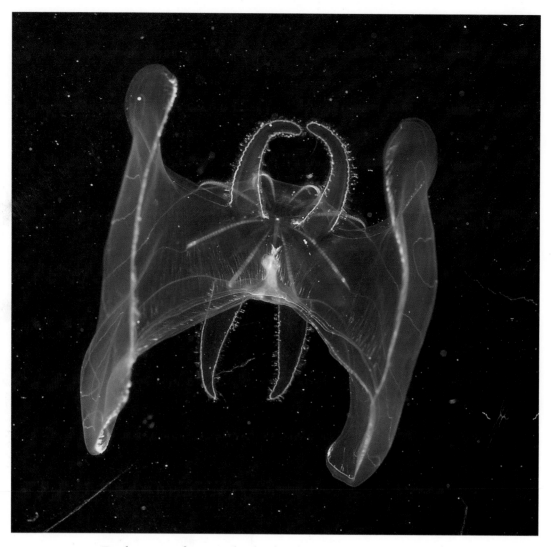

Bathocyroe fosteri *(bath-oh-sih-ROW-ee FOS-ter-eye).*

No creatures are sighted for a few moments, as the endless shower of marine snow continues. Then the scientist at the front spots a ctenophore that looks like a transparent, glow-in-the-dark bat. *Bathocyroe fosteri* is moving much more quickly than it usually does, flapping its lobes rapidly like a bird in flight. But there's no predator or prey in sight. Perhaps *Bathocyroe* is trying to get away from *Sea-Link,* in reaction to the sub's lights or the water currents created by the sub's passing.

At 2,300 feet (700 meters) a squid called *Chiroteuthis* suddenly darts up to the bubble window, attracted by the sub's lights. It looks like a rocket speeding through the night sky—a rocket with eyes. Speeding away, it is pursued by a larger and stronger squid. The second squid shoots a wad of yellow-green ink, completely covering the sub's rear window. The people inside the sub draw back instinctively, as if the ink were about to splash their faces. The ink clouds the window for a moment, then disperses. The squid probably shot the ink because it was trying to protect itself from *Sea-Link,* which it may have perceived as a predator. Many squid, when threatened, shoot out a cloud of ink that hangs in the water, seeming to mimic the squid's size and shape, while the squid itself zooms away into the darkness.

Sea-Link continues to drop. As it hovers at about 2,600 feet (800 meters), another creature drifts into view. What is it? The scientist in front doesn't recognize it, and asks the pilot to turn the sub so that the other scientist can see it, too. It looks like a new species of *Aulacoctena,* a kind of ctenophore. The scientists are very excited and talk back and forth over the intercom. Finding a new species is something most biologists never get to do, unless they spend a lot of time in a largely unexplored region such as the rain forest—or the deep ocean.

The scientist who first spotted the creature asks the pilot to try to collect the animal so that it can be studied in the lab. The pilot deftly maneuvers the sub so

Chiroteuthis *(kye-row-TOOTH-iss).*

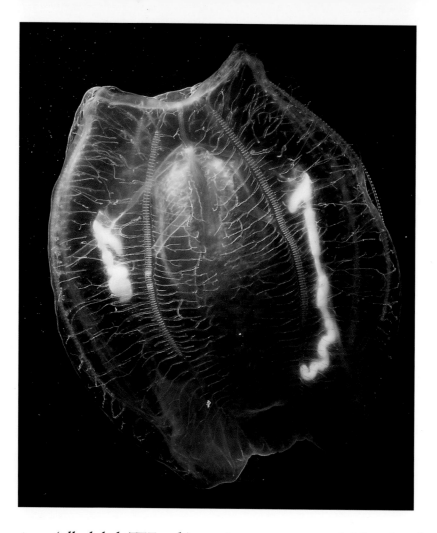

Aulacoctena (all-uh-kok-TEE-nuh), species as yet unnamed. The white lines on either side of the stomach are the animal's contracted tentacles.

that he can collect the animal with a suction sampler, a device that works something like an underwater vacuum cleaner. It consists of an oval track holding twelve containers, which have already been filled with sea water. One of these containers is in the track's "filling station," where two hoses are connected to its top; one hose leads to a water pump, the other to a transparent funnel. When the pilot has the funnel positioned next to the animal, he turns on the suction sampler's water pump. Water is sucked out of the collecting jar through one hose, while water and *Aulacoctena* are sucked in through the other. A mechanism closes the lid of the jar and *Aulacoctena* is captured alive and well. The scientists hope to be able to fill all the jars in the sampler with animals, which will later be photographed and studied in the ship's aquariums.

A siphonophore, *Bargmannia elongata,* glides elegantly past the sub a few minutes later, its bells drifting behind it like a bunch of glass goblets, its long and nearly invisible stinging tentacles trailing still further behind. The pilot maneuvers *Sea-Link* so the currents created by the sub's thrusters won't damage this delicate animal.

The scientist in the back spots a female *Phronima sedentaria*. Because she is traveling inside the hollowed-out remains of a salp, the scientist knows she is getting ready to lay her eggs. Before a female *Phronima* lays her eggs, she eats out the insides of a salp, leaving the outer barrel-shaped tissue. Then she trims, and—using an as yet unknown process—somehow alters the tissue to form a strong, firm barrel. The scientist can clearly see the *Phronima* holding onto this salp barrel with her sharp claws as she swims through the water. She will lay her eggs inside this sturdy "nest" and keep swimming with it until the eggs hatch. As well as being an inge-

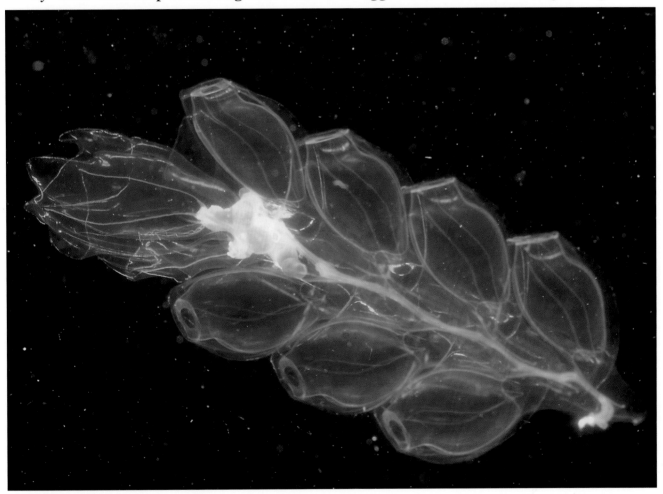

Bargmannia elongata *(barg-MAN-ee-uh ee-long-GAH-tuh)*.

41

nious mother, *Phronima* is also a midwater "commuter." This crustacean migrates vertically at dusk to feed, entering the shallower epipelagic zone. At dawn, it descends to the dark depths once again.

It's nearly time for *Sea-Link* to head back to the surface. But not before one of the deep water's most startling creatures appears at the side window. It is a viper fish, equipped by nature to survive in most extraordinary ways. It has very large jaws relative to the size of its body, which is about 12 inches (30 centimeters) long. The jaws are ringed with needle-sharp teeth that curve inward. Long, sharp, curved teeth are common in midwater fish. The late N. B. Marshall, a British marine biologist who studied these fish and other animals of the deep sea, used the term "fearsome dentition" (scary teeth) to describe this feature. Yet those recurved teeth are more than a scare tactic; they are a necessity. With so little food available, nothing once grabbed can be allowed to get away. The viper fish can push its jaw forward

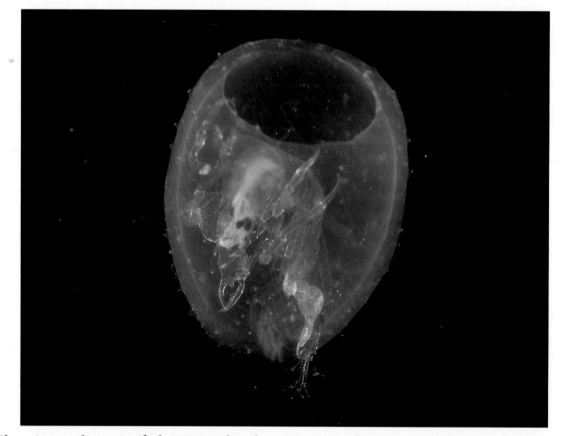

Phronima sedentaria (froh-NEE-muh sed-un-TEHR-ee-uh). This animal's genus name means "clever, thoughtful, sensible," perhaps a compliment on the female's barrel-making abilities.

A viper fish or Chauliodis sloanei *(cawl-ee-OH-diss sloan-ee-i). The photograph at the right was taken in a lab and gives a clear view of the fish's "fearsome dentition."*

so that it can get a good grip on large prey. This animal is clearly built to make the most of any opportunity to get food.

The pilot checks the air supply and the time and decides to resurface. He opens the valves that blow air into the ballast tanks. The sub starts to rise at about 100 feet (30 meters) per minute. It will maintain this rate until it reaches the surface.

The scientists sit back and enjoy their last view of this dark world. At 100 feet per minute, it will take about 30 minutes to reach the surface. It may be some time before the scientists have the chance to take another trip in *Sea-Link*. Use of a research submersible is costly and must be scheduled years in advance. But the scientists are satisfied with their day's work. With all the photographs, videotapes, and samples they have taken, they have a lot of new information to study.

The lights in the sub are turned off. At first the blackness that surrounds *Sea-Link* seems darker than nighttime, but gradually everyone's eyes adapt. They watch for the glow of bioluminescent creatures that are stirred up by the sub's passing. Sometimes the pilot flicks on the lights to see what caused a particularly bright flash, but almost always, nothing is visible but water and marine snow.

43

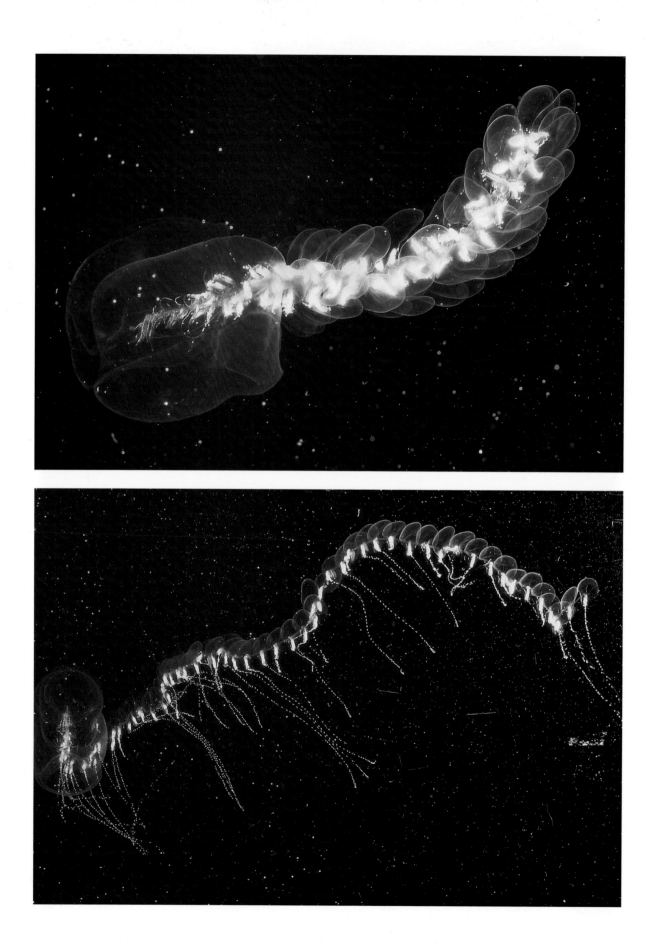

A sequence of photographs shows the Rosacea flaccida *(row-ZAY-see-uh FLAH-sid-uh)
letting down its tentacles to form a large "fishing net."*

As the light from the surface begins to filter down to the sub, the scientists spot the siphonophore *Rosacea flaccida,* a creature that is adapted ingeniously to the scarcity of food around it. The scientist takes a series of photographs that capture the *Rosacea* in the act of setting its "fishing lines." It slowly relaxes its contracted tentacles, allowing them to hang down, until they form a curtain of tentacles that allows the animal to "fish" in a much larger area than it could with its tentacles contracted. *Rosacea* will hang this way most of the day, barely moving, simply waiting for prey to come along and be trapped, paralyzed by its stinging cells.

Meanwhile, the divers must leave and allow these creatures to go about the tricky business of survival in this challenging environment. For all that they have seen today, an unimaginable amount remains unseen and undiscovered.

The Benthopelagic Zone

The benthopelagic zone is not defined by depth from the surface, as the epipelagic and mesopelagic zones are. Instead, it is defined in terms of distance from the seafloor. The animals in this zone, many species of which are found nowhere else in the ocean, live within 330 feet (100 meters) of the seafloor at depths below 450 feet (150 meters). More animals and more kinds of animals live here, feeding off the seafloor, than live in the bathypelagic zone above. With the temperature here hovering just above freezing and the water pressure up to hundreds of times greater than air pressure on land, this is a very inhospitable place for human explorers.

Until fairly recently, scientists assumed that the ocean floor was more lifeless than a desert, that nothing could survive in its endless blackness and under such enormous water pressure. But since deep-diving research submersibles came into use in the 1960s, many amazing creatures have been found.

Just how deep benthopelagic creatures exist is a question that can't be answered yet. The technology that allows scientists to explore the deep marine trenches—some of which are more than 6 miles (10 kilometers) deep—is constantly being improved and developed. Someday soon, we will reach the bottom of the last frontier.

pressure drop

More than one mile below the surface, in the total darkness and near freezing temperature of the deep sea, pulses a jellyfish with a body that looks like an Easter bonnet. The creature, called *Benthocodon pedunculata*, spends most of its time looking for food, as do all creatures who live in this harsh place.

Joining the three or four people who have ever seen this animal in its natural environment is the three-person crew of the submersible *Alvin*. Crammed into

Benthocodon pedunculata *(benth-o-CO-don pih-dun-kyoo-LAH-ta).*

Alvin *being lowered from its mother ship,* Atlantis.

Alvin's tiny passenger sphere, which is less than 7 feet (2 meters) in diameter, are the pilot, a chemist, and a biologist—who is now busy videotaping the *Benthocodon*'s search for food. Despite the discomfort of their tight quarters, the scientists are thrilled at this chance to explore a region where only *Alvin* and four other existing subs can go. For most oceanographers, a day on *Alvin* is a once-in-a-lifetime experience.

During the three decades in which deep-sea submersibles and crewless robot vehicles have been around, these craft have explored undersea mountain ranges far larger than those on land. They have plunged into slashes in the earth's crust that are as deep as 35,000 feet (11,000 meters)—deeper than Mount Everest is tall! They have traveled to the sites of undersea earthquakes and volcanoes. And they've discovered life, amazing forms of life, in places nobody ever suspected.

Alvin, which has been in operation since 1964, has seen all this and more. The sub's long and proud history of deep-sea exploration is unmatched. *Alvin* has gone where no people have gone before, and has been used to locate lost hydrogen bombs, to find sunken ships, including the famous *Titanic* and *Bismarck,* and to discover new life. Once, during a launch from her previous mother ship, *Alvin* slid off the launch platform and took on water. As the pilot and passengers scrambled out to safety, the sub filled with water and sank to the seafloor. *Alvin* was finally recovered after ten months. It was fine, as was the bologna sandwich left in it by a crew member. Preserved by a combination of high pressure and cold temperature, the sandwich was still good enough to eat.

In the 1970s, *Alvin* was exploring a hydrothermal vent, a crack in the seafloor that sends up clouds of poisonous chemicals and water heated by the earth's core. Nobody imagined that anything could live in such a place, where the temperature is as hot as 650° Fahrenheit (around 350° Celsius) and the chemicals coming out of the vents are highly toxic to almost all known forms of life. Yet, incredibly, the area around the vent turned out to be an oasis of life, home to tubeworms as tall as a person, clams the size of dinner plates, and an entire menagerie of never-before-seen animals. Unlike most of the living things on our planet, these animals do not depend on the sun's energy for life, nor on the prey that supports the creatures of the sunless mesopelagic zone. They get their life energy directly from the toxic stew of chemicals that shoots out of the earth's crust.

If such an unimaginable community can exist and thrive, what other new life forms and environments can be found on the seafloor? Each submersible dive offers a tantalizing opportunity to glimpse the unknown. But these glimpses are all too brief—five hours is all the time the crew of *Alvin* has for their mission in the deep waters—and all too rare. It's incredibly hard to reach this world!

This day's dive has been in the works for four years and has required the efforts of many people. The success of the dive today depends not only on *Alvin*'s pilot and crew, but also on the many people aboard *Alvin*'s mother ship, *Atlantis*, which carried the submersible to this dive site. These include a crew of specially trained *Alvin* technicians, divers who help launch and retrieve the sub, the crew who operate the mother

Alvin *begins its descent.*

ship, and other scientists waiting for their chance to dive, as well as backup technical and scientific crew.

The scientists stay alert to all that is happening outside *Alvin's* three tiny portholes. To see anything, the biologist has to push his face right up close to the Plexiglas window beside his seat. A tape recorder hangs by a chain nearby, ready to record any of his observations. He holds his own still camera and video camera on his lap. There's also a reading light, a flashlight, and a monitor displaying depth and time. The chemist has the same setup on the other side of the sub, just an arm's reach away.

Outside *Alvin's* window, more than twelve lights are mounted. A 35-millimeter camera that takes in a panoramic view of the area around the sub is mounted on *Alvin's* prow, along with a video camera specially adapted to record images in low light. A color video camera on *Alvin's* starboard side records close-up views. The two scientists can see images recorded by all three cameras on video monitors.

It's chilly inside *Alvin,* about 40° F (5° C), so everyone is dressed warmly. It's taken two hours for the sub to reach the seafloor, where the crew can begin the

The tiny styrofoam cup at right, which was tied to the outside of Alvin *before a dive, is a favorite souvenir of those lucky enough to journey in the sub. The water pressure the sub endures is so great that it forces every molecule of air out, miniaturizing the cup.*

Deep-Sea Diversity:
The Rain Forest of the Ocean

When studying what lives in different environments, biologists have found that the less food available, the less biomass—the amount of all living things in the area—will be found. Diversity, the number of different species in the area, may also be less.

Food is most abundant in the sun-filled epipelagic zone, and it is in this part of the ocean that most of the sea's biomass is found, and marine life is most diverse. There are fewer animals, both in number and in type, in the mesopelagic zone.

Below the mesopelagic zone, at depths below 3,300 feet (1,000 meters), is the bathypelagic zone. This zone is rarely visited by biologists in submersibles, both because there are very few subs that can go this deep, and because there is probably very little

Two creatures which may be encountered in the rarely visited bathypelagic zone, the fangtooth(*Anoplogaster* [an-OP-lo-gas-ter]), left, and the gulper eel (*Saccopharynx* [sack-o-FAHR-inx]), which can be as long as five feet.

to see. It is entirely dark here, and the temperature is just above freezing. Food is very scarce. By studying the catch in sample nets, which are best able to collect fish and crustaceans as opposed to the more delicate invertebrates, scientists estimate that there are only one-sixth as many fish in the bathypelagic zone as in the mesopelagic zone.

The trend of decreasing diversity does not continue all the way to the ocean floor, however. Oceanographers from Woods Hole Oceanographic Institution and Rutgers University have recently shown that the deep-sea floor has enough different species within a small area to rival the diversity of a rain forest. Though there are not many of each type of animal, a surprisingly high number of different types of creatures are burrowing, crawling, digging, slithering, and creeping through the sediment that has collected on the seafloor. While our understanding of this habitat is still very limited, it seems that there is far more biomass and more diversity here than in the bathypelagic zone above. Evolution has found many remarkable ways to fill the environment that exists between earth and ocean, mud and water. And we are only just beginning the process of discovering them.

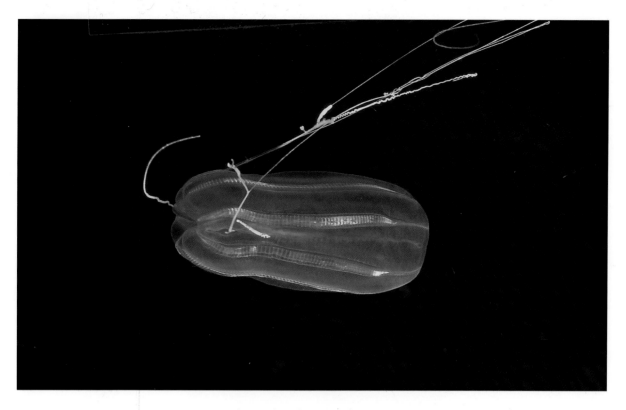

A ctenophore that has not been named yet.

first of their experiments. The chemist has had sediment traps carried down by *Alvin*. She wants to collect the particles that have fallen to the seafloor, to analyze them for their chemical components and to measure the rate at which they fall. The sub pilot uses *Alvin*'s robot arm to position the traps. The pilot has a lot of experience in working *Alvin*'s special features, and soon he has the traps skillfully positioned on the ocean floor. *Alvin*, whose location is being recorded by *Atlantis*, will return to this site on a later dive to retrieve the traps.

Meanwhile, the biologist peers into the blackness, hoping to spot another creature. He looks down at the seafloor. It looks sandy, but not like a sandy beach; the grains are rough rather than fine. The floor isn't really made up of sand at all, but of the sediment and marine snow that has drifted down from the waters above. Nothing appears to be moving here, either.

Suddenly, something catches his eye. He picks up the camera in time to get a picture of a ctenophore the size and shape of a football coming out of the blackness. As with many animals of the mesopelagic zone above, this one appears bright red in the sub's lights, though at this depth it would appear black to any predator using bioluminescence to illuminate the area. It drags its tentacles over the sedi-

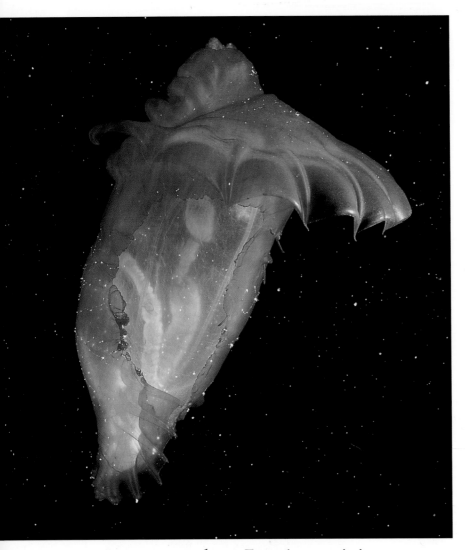

Deep-sea cucumber or Enypniastes eximia
(eh-nip-nee-AHS-tees ex-IH-mee-uh).

ment in search of food. Finding a small crustacean, it entangles its prey with a sticky substance secreted by its tentacles and draws it up to its mouth. The biologist feels lucky to be able to videotape this event, as this animal probably gets a chance to eat only rarely. The deep sea is a cold, unchanging environment with sparse food supplies. Animals that live here can't waste any energy and must live at a very slow pace.

No one knows how old these animals are, or why the jelly-like animals of the deep sea are so large in comparison to their shallow-water relatives. It's believed that they take a very long time to grow to this size. They do not actively hunt. They drift along and, if they are lucky, they will eventually drift by some food. They may use a sense similar to smell, or a vibration-detection sense, to home in on their meals. Their large size also probably improves their chances of bumping into some food.

Having successfully fed, the ctenophore drifts out of range of the sub's lights, and the biologist waits patiently for the next creature to come into view. But it is the chemist who spots a deep-sea cucumber flying over the seafloor, propelled by a waving collar of fluttering, fused tube feet. As it passes beneath, *Alvin* continues to hunt for creatures.

The pilot slowly cruises along the bottom with all the sub's lights on. Now it's his turn to spot something, an animal he nicknames "Dumbo." He quickly maneuvers the sub so the animal comes into the biologist's view. There's no denying that the little octopus does look for all the world like the cartoon elephant. The scientist films "Dumbo" as it flutters down to the seafloor, coming to rest in the sediment, its tentacles curved around it like spit curls.

Alvin continues to move slowly over the seafloor. Nothing much other than a few flakes of marine snow can be seen for a time. Then a creature that looks like a fried egg is suddenly illuminated by the sub's lights. It's a siphonophore, but not one the biologist has ever seen before. He studies the animal closely. The "yolk" is probably a gas-filled float that the creature uses to raise or lower itself in the water. Its body is ringed by gelatinous bells, some of which it uses for swimming, some

The octopus Opisthoteuthis agassizii
(oh-piss-tho-TOOTH-iss ah-gah-SEE-zee-eye).

The "fried egg" siphonophore, Stephalia corona
(steh-FAL-ee-uh-cor-OWN-uh).

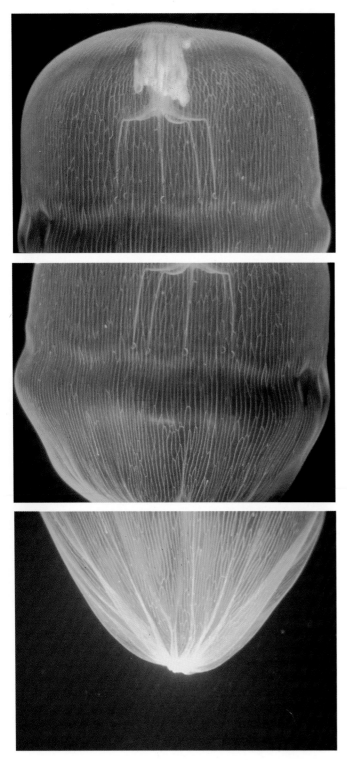

It takes three pictures to show the size of Deepstaria enigmatica *(deep-STAR-ee-uh en-ig-MAT-ik-uh). The large ripple in the middle and bottom pictures is made by contractions of the creature's muscle bands.*

for feeding. Like all siphonophores, this one has nematocysts, stinging cells, on the ends of its dangling tentacles. As the siphonophore shifts, giving the biologist a sidelong view, its body parts look like petals around a yellow-centered flower.

The biologist debates collecting the siphonophore, but decides against it. It's probably too fragile to survive the trip back to the surface. But he still has the video recording and pictures he's been taking all along, which he can later use to identify the siphonophore.

The pilot brings *Alvin* up slightly, so that it hovers at about 330 feet (100 meters) above the seafloor. The biologist, his face pressed against his porthole, is startled to see a very large jellyfish, *Deepstaria enigmatica,* rising up past the sub. It is so large that it takes several moments for its body to move past the small window. Though the biologist never gets a look at the whole animal, he can see it "flexing its muscles," as a wave of contracting muscle bands travels up through the animal's body.

This jellyfish looks and operates like a trash bag. It drifts, hunting for prey, its huge mouth open to surround anything it encounters. Once the prey is inside, *Deepstaria* contracts a sort of living "drawstring" around its mouth, trapping the prey. It seems that the jellyfish is trying to do just that with *Alvin*! But big as the jellyfish is, *Alvin* is much bigger. The creature bumps up against the sub, letting the crew get another good look at it, before it undulates away into the mysterious emptiness. The biologist is a little unsettled by this experience, but it also makes him very, very curious. Are there other jellyfish this large out there? Larger? He would love to capture one of them and study it. But, aside from the difficulty of getting any of these delicate animals to the surface alive, there's nothing on *Alvin* large enough to hold it.

Dive-time is almost up. The biologist keeps staring out the porthole, hoping to see at least one more unusual animal before the sub resurfaces. And what he sees is a creature that's only recently been seen alive: *Vampyroteuthis infernalis*. Tipped off to its presence by its tiny glowing tentacle tips, the pilot positions the sub so its lights reflect off the black-skinned relative of the octopus. The biologist has just enough time to get a photograph of the 4-inch- (10-centimeter-) long creature. Its

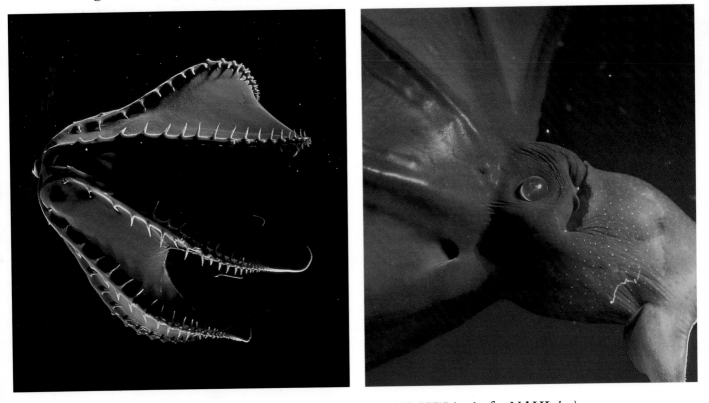

Vampyroteuthis infernalis *(vam-pie-row-TOOTH-iss in-fer-NAHL-iss).*
*With its spiked tentacles flipped over its head, left, this creature
has a strong defense against predators.*

57

tiny size doesn't seem to fits its scary name, which literally means, "vampire squid from hell."

The sub's time near the seafloor is at an end. The pilot turns off the lights and prepares to resurface. After five hours of total concentration, the scientists are very tired. As the sub surfaces, they review their notes, try to relax, try to stay warm, and don't try too hard to stay awake.

As deep as *Alvin* can dive, the deepest portions of the sea remain beyond its reach. But advances in undersea technology continue to be made. The Japanese research submersible *Shinkai 6500* can dive deeper, as far as 19,500 feet (6,500 meters), and has already begun to explore the seafloor at great depths that are unreachable by all other research subs. *Jason*, a remotely operated vehicle that can be controlled from a ship's deck, can dive for days on end, far longer than any submersible with a human crew. New self-propelled subs not connected to a research ship by a tether will be able to patrol the seafloor for days, weeks, even years at a time, while capturing video images and sending many different kinds of measurements to scientists on land. They will also be able to perform experiments and deliver the results. Who knows what creatures they will discover at these unfathomable depths?

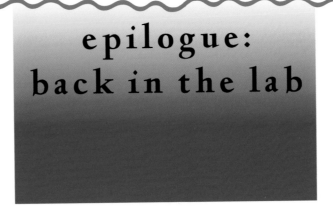

epilogue: back in the lab

Deep in the bowels of a marine research station, a biologist yanks open a heavy, battered door painted industrial green. As she enters the wet lab, she is surrounded by the sound of rushing seawater, which is being pumped in continuously from the nearby harbor. The water runs into shallow basins that contain a series of aquariums. This is the new home for some of the creatures she and others have collected at sea.

Here in the lab, the tables are turned. Now it is the humans who are in a familiar environment and the ocean creatures who must be kept in containers.

The scientists handle the delicate creatures, collected with such difficulty, with great care. To keep these creatures healthy and growing, they have recreated a small part of an ocean food chain. One tank holds a culture of phytoplankton that the biologist is growing as a food source. She checks that the phytoplankton are getting the light they need to survive, and she adds nitrogen to the water, which allows them to grow and thrive. Another tank holds tiny crustaceans that feed on the phytoplankton. These in turn are food for the small blue jellyfish in a third tank.

The scientist bends over to examine a fourth tank, which holds a chain of salps. The salps graze continuously, like cows, on the phytoplankton that the scientist adds to the tank in precise amounts. At timed intervals, she collects samples of the water surrounding the salps. She'll take the samples back to her lab to analyze them, measuring how much the salps are eating and how much nitrogen they add to the water in the tank as they digest their food.

This work is part of a long-term study she has been working on for years. She wants to understand how these tiny animals have adapted to their world and how they exist in the huge and continuously moving ocean. Though she loves her time

59

at sea, it is back in the lab that the animals she has observed and collected begin to make sense to her. The work here is painstaking and slow, but she is patient, for she recognizes the importance of trying to learn about the species we know so little about. Who knows the impact her discoveries might have?

Days spent at sea are highlights of an oceanographer's life. It is for the opportunity to experience those rare and special moments suspended deep within the "inner space" of the ocean that many oceanographers first chose this career. Yet most of an oceanographer's life is not spent on the deck of a ship. It's spent in wet labs like this one, or in dry laboratories equipped with microscopes, computers, and technical gear. It is at sea that inspiration is found. It is back in the lab that the inspiration is turned into hard data, meaningful conclusions, and, perhaps, new ways of looking at the world, through the dark blue, ever-changing window of the deep.

We shall not cease from exploration
And the end of all our exploring
Will be to arrive where we started
And know the place for the first time.

—T. S. Eliot, *Four Quartets*, "Little Gidding"

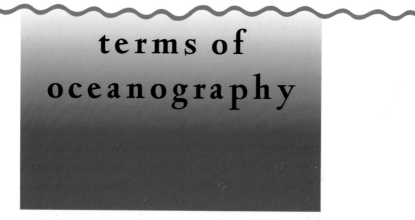

terms of oceanography

The page numbers at the ends of some definitions indicate where you can find photographs of the different types of marine animals.

Bathypelagic zone: the completely lightless section of the water column, from 3,000 feet (1,000 meters) to five miles (8,000 meters) deep.

Benthopelagic zone: the 600 feet (200 meters) of water above the seafloor in places where the seafloor is no more than 6,000 feet (2,000 meters) deep.

Bioluminescence: light produced by some living creatures.

Biomass: the amount of plant and animal life in a given area.

Blue water: the part of the ocean that is at least one mile from shore, has an average depth of 3 1/2 miles (5,500 meters), and is free from the shore's influence.

Comb jelly, *see* Ctenophore.

Countershading: a form of camouflage used by most vertebrates in the epipelagic zone that allows them to blend in with the waters around them. See page 20.

Crustaceans: a group of animals that have segmented bodies, a hard protective covering, and two pairs of antennae. This group includes all lobsters, shrimp, and crabs. See pages 17, 21, and 42.

Ctenophore: a jelly-like animal with eight "comb-plates"—groups of hairlike filaments that it uses to move through the water. Ctenophores look like jellyfish but do not have the stinging cells called nematocysts. Also called "comb jellies." See pages 19, 23, 38, 40, and 53.

Diversity: the variety of different plant and animal species in a given area.

Epilpelagic zone: the uppermost part of the water column, away from the shore and extending from the surface to a depth of about 450 feet (150 meters). All plants in the ocean grow in the sunlit waters of this zone.

Invertebrates: animals that have no backbones.

Jellyfish: saucer-shaped invertebrates with a jelly-filled body, called a "bell" or "umbrella," and tentacles with which they can sting and capture prey. See pages 4, 18, 26, 27, 36, 37, and 47.

Marine snow: plants' and animals' tiny remains that constantly fall from the upper waters of the ocean to the ocean floor.

Mesopelagic zone: the middle part of the water column, between the epipelagic and bathypelagic zones, from around 450 feet (150 meters) to around 3,000 feet (1,000 meters) deep. This is the "twilight zone" of the ocean, where there is still some light, though not enough to support plant life.

Nematocysts: cells attached to the tentacles of all jellyfish and siphonophores and used to sting or paralyze prey.

Photophores: complex light-producing organs found on some marine animals.

Photosynthesis: the process by which plants convert the energy of the sun into food.

Phytoplankton: tiny, usually microscopic marine plants.

Terms of Oceanography

Siphonophores: colonies of jellyfish-like animals that work together as a group and usually look like single creatures. See pages 25, 41, 44–45, 55 (right), and 56.

Tunicates: invertebrates closely related to vertebrates, whose bodies are covered by a membrane called a "tunic." All ocean-going tunicates have transparent tunics. See page 15.

Vertebrates: animals that have backbones.

Water Column: all the water from the surface to the seafloor. Different environments up and down the water column have different names, based on depth, temperature, the presence of sunlight, and the creatures that live there.

Zooplankton: small or fragile marine animals that move primarily by drifting through the ocean.